D0760125

DEVAS·FAIRIES AND ANGELS

A MODERN APPROACH

WILLIAM BLOOM

GOTHIC IMAGE
PUBLICATIONS

© William Bloom 1986
All rights reserved

Typeset by Wordsmiths
19 West End, Street, Somerset

Published by Gothic Image Publications,
7 High Street, Glastonbury, Somerset

Cover design by Jem Jarman 1986

ISBN 0-906362-05-9

First published 1986
Second Impression 1990
Third Impression 1992
Fourth Impression 1993

Introduction

Throughout history, in all cultures, in all major religions and in all geographical areas, men and women have talked and written about beings and creatures who belong to another magical, religious or psychic dimension. These creatures are not human beings in another dimension, but are a distinct and different life-form. They manifest themselves in a wondrous variety – from the tiny fairies who sleep in the buds of buttercups through to awesome Archangels who wield cosmic force itself. There are many names for their different types – sylphs, undines, salamanders, gnomes, elves, fairies, goblins, cherubim, seraphim, angels ... and so on. In Sanskrit there is one word which covers all these beings. This word is *Deva* which means *shining one* or *being of luminous light.*

Well, what are we to make of all this? There seem to be three possibilities:

The first is that for thousands of years storytellers, in folklore and religious text, have enjoyed simply *inventing* these beings.

The second possibility is that the human brain and human psyche are structured in such a way that regardless of time, culture or geography, people always imagine and hallucinate in the same form.

The third possibility is that devas are indeed a reality, but that they exist in a dimension normally not perceivable by the usual five human senses and, therefore, incapable of being proven by contemporary science.

The purpose of this short booklet is to affirm that devic life is indeed a reality. It attempts to present a description and an analysis of their existence which is both in tune with human logic and intuitively illuminating. Acceptance of their existence, however, is a matter of individual choice. But it seems possible that the time is ripe,on a higher turn of humanity's educational spiral, once more openly to acknowledge and to cooperate with that other life-stream with whom we are the co-creators of our existence.

This does not mean that we should enter a world of glamorous superstition or a world of naive belief. What it does mean is that we can become intelligently aware of a major reality of existence which is essential for any holistic understanding of life.

First Principles – Devic Essence

Modern science provides an interesting mystery in its investigation of the nature of dense matter. Any dense object when investigated with the use of a high electron microscope reveals itself to be made up of various electric charges and particles in constant movement. Yet, somehow or other, these electric charges, constantly moving in space, create the appearance of form. We can see and touch the dense matter of a rock, of a table or of our bodies – but place any of these materials under powerful magnification and they dissolve into a sea of moving electric charges. How these electric charges actually arrange themselves to produce *form* – the stuff we can see and touch – is a mystery to contemporary science.

Obviously there is a factor, as yet undiscovered by orthodox investigation, which *bridges* between these electric charges to create form. Unorthodox philosophy and perception suggest that this missing factor which bridges between the basic charges of atomic matter to create form is *devic essence.*

Devic essence bridges between the constantly moving electric charges to produce the matter we can see and touch. Electric charges and particles would not manifest as perceivable coherent life if devic essence did not bridge it into existence. Without devic essence there would be no coherence to manifest life – just an ocean of unconnected electric charges. *All* form is a mixture of, a relationship between, devic essence and atomic electric charges.

The devic essence of dense physical matter is the most elementary form of devic life and is the base of a natural hierarchy of devic life. This devic natural hierarchy is similar to the natural hierarchy of *atomic life* which also begins in dense matter as the *mineral* realm and which progresses *mineral* → *plant* →- *animal* → *human.* One flow of the devic natural hierarchy could be said to run: elementals → elves → cherubim → angels → archangels.

There are thus two parallel life-streams. The thrust of one of them is based in atomic matter – mineral, plant, animal and human. The thrust of the other one is based in devic essence.

What distinguishes lives of the atomic stream from lives of the devic stream is not the type body or vehicle that they have. What distinguishes between them is the *nature of the consciousness* which is incarnate in their bodies. There is, of course, a substantial difference between the form of a human body and the form of an angel or between the body of a plant and the body of fairy – but the crucial difference is in the nature of their consciousness and of the purpose of that consciousness.

Like the atomic life-stream, the devic life-stream has a natural hierarchy of expanding consciousness and awareness. In the atomic life-stream, consciousness and awareness expand from the relatively simple awareness of, for example, a rock to the complex multi-dimensional awareness of the human psyche. There is a similarly wide spectrum of difference manifest in the consciousness of the devic life-stream. For the sake of clarity in language, one might generally say that when devas reach that stage of *self-awareness* which parallels the human stage on the atomic hierarchy they become known as *angels*. Moreover, one should clearly recognise that, like the atomic heierarchy, the devic hierarchy is an *evolving* life-stream. Just as the consciousness incarnate in the mineral realm evolves to incarnate as plant, then as animal, then as human, so devic consciousness is also expanding, growing and evolving. Angels, like humans, are seeking experience and change. Moreover, of course, the evolution of both streams can pass and transcend the stage of human awareness.

The Difference between Devic and Atomic Consciousness

The relationship of devic essence to atomic electric charges in dense matter provides a metaphor for the relationship between devic life and atomic life at all levels. Devic essence 'knows' and provides the pattern by which atomic electricity can manifest in form. At all levels of life, devic life does the knowing and bridging – while atomic life does the creative doing.

Thus the crucial difference between devic and atomic consciouness is that devic consciousness is, by its very nature, *expanded* and naturally cosmically aware. Atomic consciousness, on the other hand, is by its very nature *focussed* and limited. Devic consciousness is not bound by form, but envelopes and enfolds it. Atomic awareness is limited by the form in which it is incarnate and seeks to expand awareness beyond its form. Atomic consciousness seeks to become open to, aware of and resonant with its cosmic environment. The human being, for example, seeks to expand consciousness, in order to transmute and to transcend the limitations of incarnation. Devic consciousness, however, is always aware of its cosmic environment and seeks to become more aware of that which it enfolds and envelopes. The devic and atomic life streams are interdependent in the structure of all manifest life. Their relationship is interwoven and patterned from an extremely deep level of creation.

Understanding of the differences between the two life streams may be aided by the following phrases:

Devic essence is bridging and expressive. Atomic matter is electric and dynamic.

The *dynamism* of atomic matter can only manifest in form if it is *expressed* by devic essence.

Atomic matter provides the *impulse* and *thrust* for manifestation. Devic essence provides the *matrix* for manifestation.

The atomic stream of evolution – minerals, plants, animals, humans – is oriented to *doing*. The devic stream of evolution – gnomes, fairies, angels, archangels – is oriented to *being*.

The atomic stream sees colour and hears sound. The devic stream sees sound and hears colour.

The atomic stream is seeking expansion. The devic stream is seeking focus.

The atomic stream is becoming aware of unfoldment. The devic stream is becoming focussed on creation.

The following everyday experience may also help conceptual understanding: When one whistles one vibrates out a sound wave that resonates upon one's ear drum. Yet something has given the vibration the form of sound. The atomic matter *does* the vibrating of air. The devic essence, however, bridges the vibration into the expression and *beingness* of sound. This illustration becomes even more interesting if applied to music: an orchestra is full of women and men madly vibrating air, yet those vibrations are given the form of glorious music. What gives the vibrating air the form of music? Devic essence bridges the atomic activity into manifestation.

The natural expansiveness of devic awareness, gives devas an instinctive knowing of cosmic patterns, of cosmic proportions and of cosmic harmonies. This is not a self-aware knowledge, but a natural attunement and resonance – a natural part of a deva's state of being. In its existence with manifest life, a deva does not lose contact with its cosmic source; its consciousness is openly in resonance with the essential nature of all life. It has no sense of separation from source. It can, in fact, hardly even relate to the notion that there could possibly be separation in consciousness from source. Devic consciousness, even in its most undifferentiated state, is in open attunement with cosmic essence and patterning.

In its relationship with any living form, therefore, the deva has an instinctive knowledge of the *archetype* of that form. In this context, archetype does not mean a static divinely-imposed blue-print. Nor in this context does it mean some great being upon whom all lesser beings of the same type model themselves. Archetype here means an open matrix of perfect multi-dimensional proportions within which any life may grow to its own type of perfection. There is a certain open structure which is part of the essence of any life-form – that of a diamond, or rose bush, or giraffe, or human. The deva who is involved with any particular life-form holds a multi-dimensional note around that life-form, a note which the life-form may follow and within which it may grow. This note does not constrain, but suggests the possibilities for achieving a harmonious and perfect reso- lution of that life-form's purpose. This devic note is essential for the life-form to be able to change because it provides the bridge of awareness to what may be. Just as in dense matter, devic essence *bridges* the atomic charges into manifestation – so, on a higher turn of the spiral, devic consciousness *bridges* the changes and growth of more complex life-forms into manifest- ation.

A gnome, for example, *knows* how a rock will change to crystal because it has a natural awareness of the appropriate archetypal flow – and the gnome holds the awareness which bridges the change from rock to crystal into manifestation. An elf *knows* how a bush will mature and cyclically blossom, and bridges these changes into manifestation. A cherubim is aware of how the energy should flow for a particular spiritual ceremony or ritual, and bridges that flow into manifestation. A healing angel is aware of how a healing may be perfected and bridges that into manifestation. A great Archangel is aware of how a particular atrological energy may effect a change in the human community and bridges that into manifestation.

But in all the above examples, and in life generally, the impulse for beginning the change – the impulse for *doing* – comes from the atomic existence: from the rock, the plant, the ritualist, the healer, the constellation. The atomic stream provides the impulse and the doing. The devic stream provides the bridging and manifestation.

The Body of a Deva

Although, of course, their sizes, their planes of focus, their vibrations and their purposes vary immensely, it is possible to describe the general features of a devic body.

Its core form is that of a double spiral or double vortex. Above the double vortex is its focus of consciousness. And above the focus of consciouness is a fluid, open vortex of sensitivity which, in its essence, is attuned to cosmic awareness. Extending downwards and outwards is an energy cape which may enfold the life with which the deva is working. This cape acts as a guiding matrix within which that life may unfold.

The nature of the cape and the double vortex changes as an individual deva evolves. So also does the focus of consciousness. It is the focus of consciousness, though, which retains the memory and the lessons of all past experience in a similar manner to the 'permanent atoms' of the human soul. This focus of consciousness is also directly aware of the archetypes of the fieldwithin which it is working. This focus of consciousness blends its attunement to cosmic proportions with the physical world realities of the life it is overshadowing. For example, an elf has a perfect archetypal sense of the proportions within which a particular shrub may grow, but the physical body of the plant is, in fact, affected by wind, animals, pruning and so on. The elf is thus always adjusting its sense of how its particular shrub may grow. The elf takes account of what is happening physically to its plant and, from that real physical experience, reformulates the perfect pattern of how that plant may grow within the harmonies of cosmic proportion. It is in this way that the elf, itself, also learns from experience.

The devic focus of consciousness is always synthesising its open awareness of cosmic perfection with the slings and arrows of incarnate experience. The synthesis is then held in focus in the form of a new multidimensional blue-print for the life with which it is working. And life by life, of course, the deva gains great experience, understanding how to *do* as well as how to be and learning how to achieve a point of creative focus.

The plane, or the frequency of the electromagnetic field of consciousness, in which the devic body exists depends upon the stage of evolution and the purpose of the particular deva. It is also true that very few devic beings take on a body of dense physical form for this would be too constricting and limiting of their purpose. Their vehicles, however, are frequently made up of that level of energy-matter which can sometimes be perceived as shimmers in the air, for instance around a tree, or as dancing shapes in fire-flames. Equally, a deva's size, colouring and tone will depend on its purpose. An elf may be only an inch high with the peak of its consciousness barely touching the human realm of emotion, whereas an Angel who overshadows a city or a natural magnetic centre may be hundreds of feet tall and with a consciousness that far transcends human mentality and intuition.

The natural state of devic consciousness is to be open to impression. A deva is continually openly aware of cosmic proportions and also continually openly aware of the life with which it is working. Because of this natural state of openness, devic capes and etheric bodies may take on the form *projected on to them* by active human minds. Thus, for example, over the millennia,fairies have taken on an etheric form which is semi-human in response to the projections of the human mind. Fairies are, in fact, whirls of energy. That they are 'imagined' as naive little ballet dancers, however, is symbolically appropriate even if physically incorrect. The great Archangel with its double-handed sword, massive wings and imperious gaze, is also symbolically appropriate although far from the reality of its actual form. Equally, that the devas who carry sacramental energy in religious ritual are imagined as soft and comforting cherubim has a certain metaphorical truthfulness.

Because of their open mode of consciousness and sensitivity, devas will sometimes absorb and mirror the physical actions of humans with whom they have been in close contact. This is often the case with devas of the garden and devas of ritual. There is a nice story about a priest who, whilst conducting his services, was continually clearing his throat and lifting his spectacles on and off his nose. Several small ritual devas, only a foot or so high, adopted his form and grew their own spectacles. During services they were to be seen toddling around and ceremoniously coughing and adjusting their glasses with great dignity.

Devic Existence

The way of life of a deva is very different from the way of life of any being of the atomic stream. Rocks, plants, animals and humans are all relatively physically immobile and are all also relatively blinkered in terms of their consciousness. Also, the whole mode of growth and change in the atomic stream is one of concentration, friction, tension and travail. Immense pressure, for example, turns the rock to crystal. The sprout has to burst forth from the casing of the seed. The animal labours for food and protection. And, of course, for the human, inner friction is the dynamic for growth and change.

For the deva, however, existence is very different. Freedom of movement through space and expansiveness of consciousness is natural. The awareness of perfection and of harmonious fulfilment does not have to be strived for, but is naturally known. The devic environment is experienced continually as divinely patterned and perfectly unfolding. Dense matter and form are not a source of illusion. Devic awareness dances harmoniously with the rhythms and cycles of time and energy as they unfold. In its essence and in its normal nature, devic consciousness is open, pure, expansive, serene, rhythmic, musical, trusting, sympathetic, empathetic, cooperative, innocent and caring. Through time, devas learn to become more focussed, more capable of taking the initiative, more positively active and more creative. Their most fruitful source of learning, then, is with human beings – a species not widely known for its sensitive cosmic awareness, but famous for *doing*.

The relationship between devas and humans is, therefore, mutually beneficial. Devas acquire an experience of focussed awareness and activity. Humans, on the other hand, surrender gently to the quiet of devic awareness and sensitivity. If this relationship is correctly achieved, then one is beginning to create the most perfect ecology. The first key, therefore, to coming into relationship with devas is actually to begin to surrender to their attributes. This means, as best one can, to begin to surrender to silence, to harmony, to calm acceptance, to knowledge that all is perfect in the cosmos. This requires a simple and continuous discipline.

Perceiving Devas

In attempting to open oneself up to an awareness of the devic reality, two major problems immediately arise. The first is associated with the attitude that one is simply not the kind of person who can ever possibly switch on this type of awareness. The second problem is associated with the very real concern that even when one does begin receiving impressions of devic reality, it is in fact one's own mind and imagination which are doing the creating and projecting – that the perception and contact are not real.

The key to switching on one's sensitivity to devic realities – after, of course, one has attuned to that reality by surrendering to their attributes – is to *give free rein to the playfulness of one's imagination, yet at the same time holding an inner stance of mental detachment.* Watch one's imagination running free. This, however, is of little use unless one is in an environment in which there is a strong devic presence. The two most easily accessible environments in which devic life can be found are gardens and places in which spiritual ritual, including meditation and prayer, regularly occur. In a very easy manner, the two types of environment can be brought together if one meditates among plants.

In the midst of plant life or in an environment of spiritual ritual, one may release one's poetic imagination to play with the *impressions* that it receives. If one works in this way on a regular, at least daily, basis then over time one becomes aware of impressions which endure and which do not seem simply to be personal projections or imaginations. Unless one has completely independent corroboration from another witness, the only way to be relatively certain of the accuracy of one's impressions is to hold them over a long period of time, at least a year, and to see if they then still 'feel' real.

I also wish to be clear about what I mean by 'impression.' I use this word carefully in order to distinguish it from any idea that one will perceive devic reality in technicolour pictures. Clear perception of devic life does not come in pictures, but in an intuited sense and impression. It is a sense that is as different from sight as touch is from hearing. It is a form of immediate knowing. It may be that with this impression one may also catch an impression of colour and shape, but the sense of colour and shape is secondary to the direct knowing. There are a very few people who have an innate visual clairvoyance, but this is extremely rare. There are, however, techniques for improving the quality and the accuracy of one's impression and I give an exercise for this as an appendix in this booklet. Accuracy and one hundred per cent clarity of perception, however, are not required in order for one to be able to work cooperatively with devic life. All that is required is the right personality attitude. And then, over time, greater certainty and greater clarity of perception grow.

Working with Devas

Further on in this essay, I shall describe specific types of deva with whom people often work, but for the moment it is necessary to discuss the two most important general points concerning cooperation:

Attitude. Whatever work it is that one is doing with devas, one must first consciously acknowledge that devas are present and that the work is facilitated by their presence. This means that one must have an attitude of belief which works, not in the future, but in the present. One does not state, "Devas *will* help with this work." One states, "Devas are with me here and now. I am cooperating with them and they are facilitating the work." It is this intentional awareness which sounds out the note to draw devas to work cooperatively with us. This awareness is not, however, one of intense focus and concentration; it is a gentle statement of recognition. One's imagination softly flows into devic reality and one's mind, careful and understanding, watches oneself. It is helpful to address them with spoken words: "Hello, devas. Thank you for being with me."

Action. For a human being, to cooperate means actually to do something. This doing, however, will be in tune with devic awareness which will bridge the intention into manifestation. The key to what action one takes is a willingness to act upon impression; but this willingness should be based in a loving attunemnt to one's work. Whatever it is that one is doing, one's approach should be calm and silently loving. One's attitude is quiet love and one's body should not radiate energy. Any attitude of power or ego will be unhelpful to this work. Having attuned to one's work and its purpose, let the mind be silent and sensitive to the first impulse, instinct or impression. Then act upon this impression. If there is doubt about the correctness of the impression and the action suggested, then one's discrimination can be based in whether or not the impression is felt to be resonant with *unconditional love*. One may also go within oneself and ask of one's point of inner silence: *yes or no?* Usually, a clear answer will respond. If one is unable to achieve a state of confidence about the correctness of the impressions one is receiving, then it is best to pause and review the state of one's own personality, particularly any ambition or impatience or any lack of confidence.

The purpose of this whole approach is to introduce human consciousness to that devic awareness of the completion of any work within the framework of multidimensional cosmic harmonies and proportions. Our elf, for example, knows the potential perfect completion of the growth of its shrub. By attuning to devic consciousness, the human gardener can prune in a manner that is resonant with the elf's knowledge of the multidimensional blue-print. Equally, the human healer in tune with the healing angel's awareness of perfect completion directs the flow and quality of healing energy with a sense of a completed healing.

In both these examples, however, the active human is also free to work from the basis of her or his own knowledge and plans. The gardener may wish to train a plant in a particular way. The healer may be aware of a patient's particular personality difficulty which requires attention in a certain way. The human is free to act creatively. At the same time, the deva is also absorbing new awarenesses and experience from working in such a way with people.

The Evolution of Devas

The natural hierarchy of devas is as varied and as wide as the natural hierarchy of the atomic stream of evolution. It ranges from the undifferentiated essence of dense matter through to glorious cosmic Archangels. Through the passage of time, experience leads devic consciousness along its life-stream and into increasingly sophisticated states of being. In this section, the principles and some distinctive features of this stream are briefly described.

Within any field of matter or any field of energy – for example, rock or water or fire – there are devic beings who are aware of the way in which those elemental forms should evolve and be patterned. (There are also, of course, elementals and devic essence which make up other energy field forms such as emotion and thought. They are beyond the scope of this booklet, but are fully described in the book *A Treatise on Cosmic Fire*; see bibliography.) These beings are, of course, familiar as salamanders, undines, sylphs and so on, and have evolved from the undifferentiated essence in which they were previously totally diffuse consciousness.

Through time and experience, they evolve a focus of consciousness and become elementally aware of the form field for which they care. These small beings also have older relatives who perform similar tasks, but on a far greater scale. There are, for example, great devas of the earth who hold an awareness of mineral and rock changes over many square miles. There are also great earth devas who attend the leys beneath the ground, the plate lines and other features of the Earth's crust. These beings have sometimes been interpreted in folklore as dragons and great worms. There are great devas of the air who are concerned with atmosphere and great movements of weather. These beings have been imagined, for instance, as great puffing faces creating winds, gales and hurricanes, and as the beings whose tension is released in thunder and lightning. The small elementals do not evolve directly into these great beings, but pass through stages and degrees of responsibility which give them ever increasing experience, sophistication and focus.

Then there is the whole range of devas who look after plant life and landscape; these have evolved from earth elementals whose environment was particularly associated with plant growth. These devas range from tiny fairies through to great angels who are concerned with the ecology of whole geographical regions. The fairies and elves who tend individual plants evolve until, for example, they are associated with individual trees. Tending an oak tree or a great pine over several centuries gives a deva immense experience of the ecology surrounding that tree. That tree deva may then pass on to responsibility for a particular copse or wood where it holds the focus and awareness of the perfect ecological pattern for that particular area.

Any landscaping, agricultural or building work done by human beings should ideally take place in harmony with the vision and knowledge of the local landscape deva or angel. However, whether or not human workers act consciously in cooperation with landscape devas, in these situations devas may begin to become very accustomed to human beings, and their further evolution may be intimately entwined with human existence. Equally, certain devas may become very involved with local animal life and their further evolution be along the stream of fauns and the beings associated with Pan. Some of the plant devas may remain working with the plant species with which they began and become responsible for the evolution of the whole genus. Others, of course, may evolve from caring for landscape in three-dimensional geography to the tending of increasingly subtle energies in multi-dimensional inner space. Sooner or later, however, because of humanity's key place in the scheme of planet Earth's natural hierarchy, all devas pass through a stage, at one level or another, in which their major focus of work is associated with the human stream of evolution.

What should be clear by now is that an individual deva, like an individual human, has its own unique history and experience which gives it its own area of speciality. One human being, for example, may have an incarnational history whose tendency is towards service through teaching or physically caring for others. Another human being may have an incarnational history of action or of working in science. Equally, a deva has its own particular past which has led it to its particular focus. Because, however, of the nature of devic existence and consciousness, as devas evolve from life-form to life-form, they do not lose consciousness and, like humans, 'die.' They simply glide on with no break in their continuity of awareness.

The booklet now goes on to describe in greater detail the activities and work of certain distinct types of deva and how humans may cooperate with them.

Plant Devas, Landscape Angels and Human Cooperation

Plant devas are known in folklore as elves, fairies and so on. These vortical creatures of dancing energy possess awareness of the blue-print of how a particular plant may perfectly evolve. This is not a static map, but more like a piece of ongoing improvised music. The actual phsyical structure is subservient to the note of beauty and the quality of fulfilment for the plant. The deva has an exact sense of what the perfect plant may be. The changes wrought by the interference of weather, of other plants, of soil conditions, of animals and of people, are all inputs into the general gestalt to which the deva fluidly adjusts – always clearly holding the sense of the perfectly fulfilled plant. Thus, no matter what external interference there might be, the plant itself is still enveloped by the ideational matrix, held by the deva, into which it may grow.

The range of awareness and of activity of the devas who do this kind of work is enormous. There are tiny trainee devas whose focus is purely the petaling of a daisy. Then there is the fairy who encapes a whole tulip. Or an elf who cares for a complete rose bush. Over time the experience and awareness of these creatures expand to encompass ever larger and more complex plant forms: a field of crops, a copse, a forest and so on. As they evolve and their experience increases, their consciousness and knowledge expand to include other crucial factors of the landscape in terms of the total ecological relationship between climate, soil, plants, animals and humans. What these devas have – and it is a feature of all devic life – is a natural sense of cosmic proportion and harmony which can accomodate all happenings and events. This sense of cosmic proportion and harmony works not only in three-dimensional space, but also in the dimensions of time, of colour and of transcendent consciousness.

In all their work they hold a knowledge of perfect pattern and of perfect process which knows the whole story, past and potential, in a time-transcending now. The actual physical plant or landscape, however, is limited by and must pass through *time* in order to grow into and to go through the pattern and the process already known by the deva in its transcendent knowledge.

If we accept that devic awareness, or devic knowing, has a perfect grasp of how a plant or ecology may perfectly evolve, then the human gardener may attempt to work with this same awareness. This does not, however, mean that the gardener or landscaper surrenders passively to what is already happening; if she does this, then she is simply meditating and not actually gardening – and the garden then becomes wilderness which is, of course, equally fine. The active gardener, however, works towards a creative purpose in terms of crop or of beauty or of playing artistically with ecology.

For the active gardener decisions have to be made about positioning, soil conditions, pruning and weeding. Each decision can be made after a few seconds attunement to the work: "Hello, devas of the garden. I am about to garden here and my intention is to work with love and harmony – aided by your awareness. I am open to your impression. Thank you for your help and for your company." Then make your decision. You may feel a sense of disapproval from the garden fairies; if so, you must then decide whether to proceed. I know some sensitive gardeners who have been unable to continue at all because of their continually sensing devic disapproval when, in fact, the fairies and elves had merely adopted the human characteristic of disapproval for experimental fun. Some of the fairies were quite perplexed by the gardener's inertia; others were giggling uncontrollably. *When working with devas, sooner or later you will have to make a decision.* Humans learn from experience, so this is fine and not to be a source of anxiety. The devas also learn from experience and they specificaly learn creativity and the possibility of alternatives from their co-work with humans.

Apart from the local fairies and elves, there are other more sophisticated devas who are associated with a particular plant species. Thus, for example, the gardener or horticulturalist who is working towards a new strain of plant may telepathically call upon the aid of the Angel whose job it is to care for that particular species and the gardener may then work within the aura of that Angel and in resonance with the plant's archetype.

Holding the plant with which you wish to work, close your eyes: "I am one with this rose and through it I am in rapport and communication with the Angel of roses. In love and harmony, I seek to work with you. I ask gently for your help. I recognise and acknowledge your presence and I thank you."

Equally, when landscaping, survey the area where you wish to work, close your eyes: "I am one with this landscape and through it I am in rapport and in communication with the Angel of Landscape and the Angel of this Ecology. Angel of Landscape and Angel of this Ecology, be with me as I work in love. I recognise and acknowledge your presence and I thank you." The wording used in this address, and in the others throughout the booklet, is suggestive and to be altered and built upon according to each person's own discretion.

This kind of cooperation not only allows the human worker to be attuned to devic awareness, but also allows the particular local devas to become aware of what the human gardener has in mind. If a gardener's or landscaper's idea is resonant with divine proportion, then the deva will facilitate its completion by holding a clear energy picture of how it should be. This is to say that the deva will alter the blue-print or archetype to allow for the new human idea and the deva will hold this new blue-print over the plant or area where the work is taking place. This can be exceptionally helpful, for example, in pruning or in the creation of a new type or strain of plant. Often a plant or landscape angel may be cooperating with several people who, unknown to each other, are working on similar projects. The angel may thus synthesise the best of their individual inputs to produce a single blue-print to inspire them all; in this way several people may simultaneously produce a new plant strain of the same type. This type of synchronicity happens in any area of research which involves angelic cooperation; coincidence of inventionis a well known phenomenon in all fields of innovation and not simply due to telepathic rapport.

It is also appropriate in this section to mention those angels who look after and work with the magnetic centres and power points across the Earth's surface. In these places, the most evolved of landscape Angels work harmonising and resonating the energies which are drawn down to and radiated from the centre. Cooperation with these angels is done in sacred ritual and in silent attunement.

At particular times of the year however – at the solstices, equinoxes and Celtic fire festivals – the major form of cooperation with the plant devas and great ecology angels is by celebration. These devas and angels have an annual rhythm of awakening, activity, intense growth, relaxation and then silence which were recognised by the great 'pagan' and fire festivals. When we dance, commune with nature and the Earth, and have celebration at the time of these festivals, we co-create a joyful communion with devic life. The time of these festivals, however, is not rigidly fixed but varies, of course, from region to region with the differing local ecological conditions.

Devas of Meditation, Religion, Ceremony and Ritual

The history of some devas leads them into a particular type of relationship with human life. It may begin, for example, when an elf finds itself tending a garden near to where a human being often prays or meditates. The person who prays or meditates has an attitude which is in tune with her most deep and essential purpose. Her attitude is detached from the short-term involvements of daily life and is attuned to the ultimate and perfect completion of the human process – an individual who is complete love. In prayer and meditation, the individual becomes separated from the illusions of temporary life and attempts to bridge into the future state of perfection. This action and this intent are similar in resonance to the plant shoot bursting forth from seed. This attitude holds a purity of purpose that is in complete harmony with cosmic perfection. The individual who prays or meditates, therefore, radiates a particular multidimensional note which a deva can sympathetically recognise. Because of the natural expanded consciousness of devas, a deva can be aware of the energy state and atmosphere which the human wishes to enter and, aware of this blue-print of perfection, the deva may help bridge the human into this new resonance.

An elf, then, who has spent much time in the vicinity of a praying or meditating human may be drawn to a future which is associated with people and not with plants. Similarly, the plant and landscape devas who are near a monastery, church, temple, mosque, convent, sanctuary or meditation area, may well be attracted by the multidimensional currents of devotion and aspiration, and may also be drawn into that line of evolution which is particularly associated with human beings. They, then, join a huge school of angels who specifically cooperate with human spiritual service and aspiration. This school of angels has mainly two spheres of interest: first, to help individuals with their private spiritual growth and, second, to help with the construction and flow of the helpful energies which are associated with meditation, prayer and ritual as forms of service. Those devas who aid individual spiritual growth are sometimes known as 'guardian angels' and those devas who work with currents of energy are the seraphim, cherubim and so on who are, for example, invoked in the Christian Eucharist.

Guardian Angels
When spiritual growth and service become the predominant theme in an individual's life and are anchored into that person's daily rhythm, then devas of spiritual growth will begin to be attracted to that individual. They work in the atmosphere around the individual helping to keep it clean and magnetically pure; they are similar to the salamanders of fire or undines of air. They block the entrance of unwanted vibrations and keep the atmosphere spiritually playful and sympathetically helpful to the individual. While they are of service to the human aspirant, they themselves are also learning from the whole experience. In particular, they learn to appreciate and harmonise with human eccentricities, especially the human ability to oscillate between an attitude of genuine spiritual sincerity and an attitude of undisciplined purposeless egoism.

When an individual's major life focus becomes personal change and unfoldment on the path to unconditional love, then a particularly experienced deva of human spiritual growth may become assigned to that individual. This Guardian Angel holds a clear multidimensional awareness of how its human can best evolve. It holds an inner picture of the individual's next spiritual step. This awareness is not concerned simply with apparent action and attitude, but is concerned with the subtleties of building the inner 'rainbow bridge', or path, to higher consciousness; it is also concerned with the subtleties of the relationship and resonances between the various chakras and energy centres of the human body.

It is possible for one Guardian Angel to look after several people simultaneously or even a whole group. Contrarily, in some cases one individual may be aided by several Guardian Angels who are all gaining useful experience from that particular person's idiosyncratic growth rhythms. In the Western mystery tradition there are rituals whose specific purpose is to achieve contact with one's Guardian Angel. These rituals are not, however, geared to a direct apprehension and perception of the Angel, but are, in fact, oriented to bringing the individual into greater and purer resonance with the *awareness* held by the Angel. In a very real sense one's own soul, or silent inner self, also holds this vision and in some symbolic language 'Guardian Angel' is actually a reference to the human soul.

The kind of consciousness which is required for full communion with one's devic Guardian Angel is one that in Buddhism might be called a higher form of samadhi and in the West might be described as "a sensible and knowing bliss." The devic Guardian Angel may also hold on to advice which is sent to an aspirant from her spiritual mentors; this advice might concern, for example, the type of meditation focus or a link that it would be helpful to make. The Angel holds this advice in the aspirant's aura until the physical brain finally anchors it. These Angels also, of course, provide a form of atmospheric protection to aid one through life's trials and tribulations, although no help is ever provided that transcends any individual's own karmic state.

Ceremony Devas
These are the beings who facilitate spiritual ceremony and ritual. They understand that the object of spiritual ceremony is to invoke and to attract helpful energies for the participants and for the environment. They understand that the meditator or celebrant – whether or not s/he is conscious of it – is creating an energy channel through which can flow energies such as pure enlightenment, unconditional love and clear spiritual purpose. It is the work of these devas to help construct this energy channel and to enhance and to increase the grace and energy that flow. These devas are of varying sophistication and are recognised in the call of the Christian Eucharist: "Therefore with Angels and Archangels, with Thrones, Dominations, Princedoms, Virtues, Powers, with Cherubim and Seraphim and with all the company of Heaven, we laud and praise . . ." C.W. Leadbeater in particular has written interestingly on the subject of these devas of ritual.

These ceremony devas, therefore, work to bridge the spiritual intent and aspiration of the ritualist into active manifestation. They help to create the contact with the source of the blessing or energy which is being invoked; and they help to build the multidimensional energy structures which amplify and enhance the flow and radiation of the energy and grace. If an individual always places within her/his meditation a time of deliberate channeling and radiation of helpful energy, then a deva such as a cherubim will always be there to help with this silent work.

There are furthermore devas who are associated and work with the energy of particular planets, particular astrological constellations and particular rays. There is a western tradition of ritual work the purpose of which is to experience and to communicate with these particular angels and the energies which these angels tend.

Healing Angels

There is a large school of Angels whose major focus is healing work. They do not themselves do dramatic acts of healing, but they provide an atmosphere of clarity, of comfort and of love, in which healing can take place. They are also capable of facilitating healing by directly helping the work of individual human healers.

This angelic help is available in all healing situations, but its effectiveness depends upon whether the angelic assistance has been *consciously* sought and invoked. Nevertheless, almost without exception, all hospitals and places of convalescence are overshadowed by powerful healing angels. Without this angelic overshadowing many urban hospitals in particular would be less effective and with less of that atmosphere which aids nurses, doctors and other staff in their work. Angelic healing is as much a reality in the high-technology hospital intensive care unit as in the spiritual healer's private ashram of care.

All healers, especially those using what are considered to be alternative methods, can usefully use angelic help in a self-conscious way. The specialist healing angel has an awareness, situation by situation, of the necessary process for the perfect healing; and the alternative healer can attune herself to this awareness. Before any session, a healer may close her eyes and surrender to an atmosphere of peaceful love: "I am one with the nature of all healing. I recognise and am grateful for the presence of this healing angel. I work with your guidance." It is possible then to attune to the awareness of the Angel and to understand the nature of the perfect healing. The healer can then *act* accordingly.

More generally, attunement to a healing angel will invoke help to cleanse and purify the atmosphere both before and after healing. This kind of help can be used not only in physical healing, but also in the psychological healing associated with transpersonal techniques.

Some healing angels become very experienced and sophisticated. They may then work with large groups of healers to develop new techniques and methods of healing work. Healing

27

Angels have also worked to facilitate the discovery of certain useful drugs such as antibiotics, anaesthetics and analgesics.

Angels of the City, Angels of the Home and Muses

In this final section on particular types of deva, I want to focus mainly on those angels associated with cities, with the home and with artistic inspiration.

Angels of the City.
There are devas with great experience of the various facets of life on our planet whose work it is to overshadow great cities. Their consciousnesses can comprehend the vast complicated ecology of modern urban centres and they provide careful nurture in several ways. First, they ensure a helpful flow of prana into the city at dawn and in the early hours of the morning so that all the plants and plant devas are kept vibrant despite the pollution and chaotic vibrations. Second, they provide a protective aura for the other thousands of devas whose work keeps them in an urban environment. Third, they keep an overview on the general state of spiritual aspirants in the city and do their best to facilitate a generally helpful atmosphere. Finally, they are aware that cities are crucially important centres in the general planetary structure and they link with other cities to form a network of consciousness in the greater multidimensional body of the planet.

In ancient times, all cities had their deity to whom offerings and respect were given. One of the most famous of these was Athena who overshadowed the flowering of Athenian civilisation. These deities were, in fact, the overshadowing Angels of the city. Acknowledgement of their presence and inspiration opened the citizens to the angelic aid that was available; the essence of worshipping these angels was certainly not a crude superstition – although like all human activities it was capable of distortion and exploitation. The temples to the city deity – the overshadowing Angel – were the simple acknowledgement of a real and helpful truth. Today their presence is not recognised, but they are in fact still with us quietly and lovingly fulfilling their work. Go at dawn to any important hill or place of worship in a city. As the first light of day begins to break, gently and silently attune to the blessing of the overshadowing Angel.

Angels of the Household.
Just as there were shrines to the city deity, so in the past many homes had their own household deity – this is, in fact, the recognition of an overshadowing angel of the home. Today, one can still have an overshadowing angel who tends to the home's atmosphere, adjusting energies so as to aid a more attuned and beautiful life. A conscious effort, however, must be made to invoke the help and ongoing presence of an angel of this kind. First, one must be aware for several weeks of one's intention to call such an angel to one's home. Second, one's home must be thoroughly cleaned. Then at a quiet and serene moment sit and light a candle for the angel. Remain quiet and sense that an angelic presence may come to the home.

"Angel of the house, I recognise you and thank you for your presence. I dedicate this house to the purpose of spirit and to the purpose of unconditional love – and I humbly ask your help in this work. Let it be a house of love and creativity aided by your beautiful presence." You may feel like burying a crystal beside your house or performing some other quiet ceremony which is attuned to your household angel. It is good to light a candle specifically dedicated to your household angel at least once a week. It is also helpful and supportive to light a candle for the angel of your city – or the angel of your surrounding landscape. The smaller fairies and elves also like little candle ceremonies. These candle ceremonies can be supplemented by other pure and symbolic offerings such as the burning of incense or the proffering of a small bowl of grain. You may find that you make a small shrine or altar dedicated to the angel of your home or your city. Children love these small gestures of sacred drama. The atmosphere in a room or a building can be settled by the most simple form of devic attunement, which is to place symbols of the four elements in four corners of the room or house: in one corner, incense for air; in another corner, salt for earth; a candle for fire; and a bowl or glass of water for water. When moving into a new home, this small ceremony of the four elements can be most useful – and it can be performed with complete casualness and no pompous decorum. If it is followed by lighting a candle for a household angel, the home will thereafter be blessed with angelic help.

Muses.

Certain devas with great experience of working with people take on the role of beings of inspiration. When an artist – whether working in paint, form, poetry or music – is seeking to re-express life in a manner that expands people's range or depth of perception, then s/he places herself in resonance with certain natural inner harmonics. A muse – a deva of inspiration – may then respond to this attunement and, aware of the essence which the artist is attempting to touch and interpret, will help bridge the artist into a more full consciousness of it. This sudden, fuller awareness may be experienced as inspiration as the artist suddenly touches and intuitively understands certain novel realities.

This understanding of the work of muses does not for one moment take away from the unique creativity of individual artists; it simply points out an inspiring devic relationship which is sometimes at work and which demonstrates in certain art. There are, for example, very interesting angels who have particular relationships with specific forms of art. There is a particular school of angels concerned with classical proportions in architecture whose common inspiration can be seen manifest in architecture of the Palladian style. Equally, there is another school concerned with the exquisite whorls and patternings of traditional Islamic art. Again, there is a school concerned with the classical spirals of choral music.

Finally, there are the muses and angels who look after particular aspects of life. At most large government buildings and places of legislature, there are angels who overshadow with archetypal ideas of open debate and just governance. There is a whole school of government devas whose source is in the Houses of Parliament in Westminster but who are now at work all across the globe. They themselves have evolved from the devas who were once concerned with tribal councils. There are also, for instance, devas of justice who overshadow courts of law. In fact, most areas of life which have an inherent dynamic of service possess great angels or muses whose work is concerned with aiding their process towards perfection. It is this same school of angels which tends the great thoughtforms and archetypal ideas on the inner planes.

As with all devas, these angels have a natural sense of how human life might be if completely harmonically proportioned, if

31

completely in resonance with the inherent perfection of all existence. They work always to hold a helpful vision and a sense of inherent perfection for their co-workers, humanity, on the planet.

Conclusion

There is devic life everywhere. This booklet began with the most basic form of devic life, the devic essence in dense matter. It touched in the following pages upon many manifestations of devic existence. It has not, however, touched upon the magnificent Angelic Beings who in their consciousnesses work with whole planes and dimensions of existence. Imagine for a moment that Angel who tends the body of the Sun or that Angel who links Sirius to our own solar system. The reality is extraordinary.

But the reality of the tiny fairy in the daisy, or the healing angel, or the overshadowing presence at a place of worship, is also extraordinary. In this short booklet we have merely skimmed across these realities. My hope is, however, that this booklet stimulates a real awareness of their presence. It is always possible to calm oneself, to centre and to be aware of this glorious and loving magic. Indeed, on a daily and ongoing basis it is worth being self-disciplined and purposeful about this. It brings personal grace and beauty, but it also crucially serves the unfolding plan of our planet— alongside the liberation of the female principle and the Green movement. Devas are inextricably part of our unfolding story, just as we are of theirs. Awareness of their existence draws us into ecological balance, a balance which works only towards greater peace and greater spirit.

May the blessing of their natural awareness of perfection draw us all closer into the unfoldment of unconditional love.

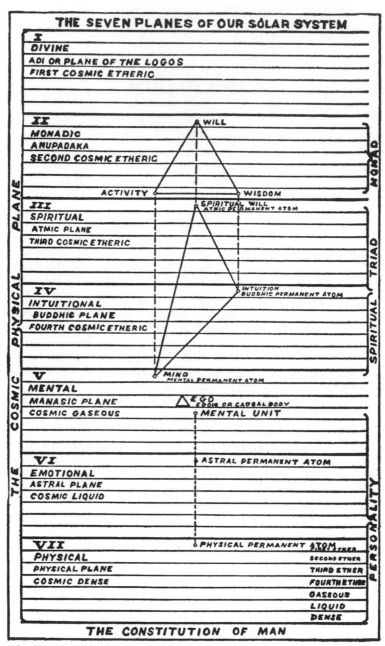

THE SEVEN PLANES OF OUR SOLAR SYSTEM

THE COSMIC PHYSICAL PLANE

| I | DIVINE | ADI OR PLANE OF THE LOGOS | FIRST COSMIC ETHERIC |

II
MONADIC
ANUPADAKA
SECOND COSMIC ETHERIC

WILL

ACTIVITY — WISDOM

III
SPIRITUAL
ATMIC PLANE
THIRD COSMIC ETHERIC

SPIRITUAL WILL
ATMIC PERMANENT ATOM

IV
INTUITIONAL
BUDDHIC PLANE
FOURTH COSMIC ETHERIC

INTUITION
BUDDHIC PERMANENT ATOM

V
MENTAL
MANASIC PLANE
COSMIC GASEOUS

MIND
MENTAL PERMANENT ATOM
EGO
EGOIC OR CAUSAL BODY
MENTAL UNIT

VI
EMOTIONAL
ASTRAL PLANE
COSMIC LIQUID

ASTRAL PERMANENT ATOM

VII
PHYSICAL
PHYSICAL PLANE
COSMIC DENSE

PHYSICAL PERMANENT ATOM
FIRST ETHER
SECOND ETHER
THIRD ETHER
FOURTH ETHER
GASEOUS
LIQUID
DENSE

MONAD

SPIRITUAL TRIAD

PERSONALITY

THE CONSTITUTION OF MAN

(Alice Bailey   *A Treatise on Cosmic Fire.*)

Appendix

What follows is a simple exercise which may help unfold a more sensitive and accurate perception of devic relities. It is an exercise which needs to be conducted twice a day, perhaps as a lead-in to one's meditation, and which requires practising for two or three years before any confidence of accuracy may be assumed.

The essence of the exercise is to attune to and to examine the quality and nature of the different planes of existence. For example, having calmed oneself and with one's back straight, one spends fifteen seconds examining in one's consciousness the nature of dense matter. What does it feel like? What is its nature? What exists in it? One then proceeds to do exactly the same for liquid. After fifteen seconds, one proceeds to the same focus upon air.

After air, one proceeds through four more dimensions of manifestation to make up a *seven*-fold set of steps. This set of *seven* is in accordance with the teachings and maps of all esoteric schools and, from personal experience, I recommend it without reservation. (If you do not resonate with the seven-fold map, you will have to rework the meditation.) Earth, air and water are the bottom three planes in the first set of seven. The next four planes are all in what is called etheric matter.

Earth is the seventh subplane.

Water is the sixth subplane.

Air is the fifth subplane.

The fourth subplane is the shimmer of energy that many people can see, for instance, around trees.

The third, second and first subplanes you must examine for yourself.

Having completed the examination of this first seven – focussing on each in turn – one then progresses into the emotional/astral plane. Again, this is divided into seven subplanes. The seventh at the bottom being the most dense. The first at the top being the most dynamic. Examine them all.

Then move on to the mental plane. Again, there are seven. On the fourth subplane one begins to move into transcendental areas of consciousness.

Then move higher still into the buddhic/intuitive plane. Again, there are seven sub-planes.

After that, there is the plane of spirit or atma.

Then the plane which is called monadic.

Finally, the plane which is called cosmic.

Twice a day progress through the planes as far as you can go, examining each one. It is obviously a good exercise just for one's own growth and expansion of consciousness. At a very practical level, though, its result is that when you do perceive something or have an impression, you will be able to identify on exactly what plane it exists and its nature. This is the beginning of a very helpful accuracy.

Select Bibliography

Nancy Arrowsmith with George Morse, *A Field Guide to the Little People*, Macmillan, 1977.

Alice Bailey, *A Treatise on Cosmic Fire*, Lucis Press,1970.

Katharine Briggs, *An Encyclopaedia of Fairies*, Pantheon, 1976.

H.K. Challoner, *Watchers of the Seven Spheres*, Routledge, 1933.

Georges Chevalier, *The Sacred Magician*, Paladin, 1974.

Gustav Davidson, *A Dictionary of Angels*, Free Press, 1967.

Findhorn Community, *The Findhorn Garden*, Turnstone, 1976.

James G. Frazer, *The Golden Bough*, Macmillan, many editions.

Billy Graham, *Angels – God's Secret Agents*, Doubleday, 1975.

Geoffrey Hodson, *Fairies at Work and at Play*, Theosophical Publishing House, 1925.

Geoffrey Hodson, *The Coming of the Angels*, Rider, 1932.

Geoffrey Hodson, *The Kingdom of the Gods*, Theosophical Publishing House, 1952.

Thomas Keightley, *The World Guide to Gnomes, Fairies, Elves and Other Little People*, G. Bell, 1878.

C.W. Leadbeater, *The Science of the Sacraments*, Theosophical Publishing House, 1937.

Dorothy Maclean, *To Hear the Angels Sing*, Turnstone Press, 1980.

Gothic Image Publications

We are a Glastonbury based imprint dedicated to publishing books and pamphlets which offer a new and radical approach to our perception of the world in which we live.

As ideas about the nature of life change, we aim to make available those new perspectives which clarify our understanding of ourselves and the Earth we share.

New Light on the Ancient Mystery of Glastonbury	John Michell	£9.95
Glastonbury - Maker of Myths	Frances Howard-Gordon	£5.50
The Glastonbury Festivals	Lynne Elstob & Anne Howes	£9.95
The Green Lady and the King of Shadows	Moyra Caldecott	£4.95
The Glastonbury Tor Maze	Geoffrey Ashe	£2.95
Labyrinths: Ancient Myths & Modern Uses	Sig Lonegren	£12.95
Spiritual Dowsing	Sig Lonegren	£6.95
Needles of Stone Revisited	Tom Graves	£8.95
Eclipse of the Sun: An investigation into Sun and Moon Myths	Janet McCrickard	£14.95
Dragons: Their History and Symbolism	Janet Hoult	£4.95
Hargreaves' New Illustrated Bestiary	Joyce Hargreaves	£10.95
Meditation in a Changing World	William Bloom	£8.50
Euphonics: A Poet's Dictionary of Sounds	John Michell	£5.95
Dowsing the Crop Circles	Ed. John Michell	£3.95

These titles are available by Mail Order. Add 20% for postage and packing - 40% for air mail to the USA and Canada.

Gothic Image Tours

If you are interested in visiting the ancient and sacred sites of Britain and Ireland, we organise tours on a regular basis. For further details contact:

> Gothic Image
> 7 High Street
> Glastonbury
> Somerset BA6 9DP
> ENGLAND
> Telephone: 0458 831453
> Fax: 0458 831666